Introduction to Metrology

Alex Lepek

Introduction to Metrology

Alex Lepek

Cover designed by the author.

To all my grandchildren.

Table of Contents

1. Introduction

This book is a brief introduction to metrology, which is the practice and science of measurement. The book addresses measurement, test, calibration, methods to estimate measurement uncertainties, quality assurance of these subjects and issues related to them at an introductory level.

The book was written considering readers who are new to the field and need a fast overview, but experienced practitioners can also benefit from the theoretical background. It does not focus on a specific type of measurement technology or measurement discipline but only on the general concepts and principles common to all measurements.

I demonstrate the principles with examples from disciplines that can be understood by intuition, such as temperature, force, length, and time interval felt by our senses. We feel the room temperature on our skin, we feel force by our muscles, we can estimate distances and lengths just by looking at them, and we can feel time interval by the amount of boredom when waiting for our dentist. In contrast, people do not feel electric fields and therefore we do not experience electric fields intuitively. Sharks and eels do feel electric fields, but this book is intended for people, so I avoid electrifying examples.

I start with an overview of metrology; next I describe important concepts at a basic level to allow for an intuitive discussion with the reader before going deeper into the principles of measurement and uncertainty estimations. The precise definitions of the terms in the book are found in the document JCGM_200_2012.pdf which is the official dictionary of metrology. It can be downloaded from the BIPM web site https://www.bipm.org. More material can be found in the package JCGM_pack_2012-10.zip which may be downloaded from the same site. Material about quality assurance in testing and calibration can be found in the ISO/IEC 17025 standard.

This book shares with the reader my point of view on metrology learned during a period of over 35 years of consulting. During this period, I wrote numerous software programs to run automatic tests and calibration procedures. Most of them are proprietary and belong to customers and will not be described here. The off-the-shelf program MetroVal was written especially for accredited calibration laboratories and complies with accreditation requirements.

This book is modest. It is not intended to be a handbook and it does not have pretension to address every issue in metrology.

Dr. Alex Lepek, Kfar-Saba, 2023

2. Metrology overview

Metrology is the practice and science of measurements. Measurements consume about 2% of the Gross Domestic Product (GDP) as reported in an estimate in 2002 [1], and therefore are an important part of the economy. In fact, the effort done in experimental physical sciences at research centers and universities is mostly measurements. Same is true for industrial development labs and a major part of quality control in manufacturing.

One of the main purposes of metrology is to make everyone agree upon the outcomes of the same measurements by different parties at separate locations. This is a necessity in national and international trade. If different parties are trading overseas and use different measurement units, they must have an agreed method to convert measurements outcomes presented in different units. Not to mention dissemination or comparison of scientific findings. It is even best if everyone uses the same units. For example, the same unit for weight. Unfortunately, even today, different countries use different units although internationally accepted units are well defined. For example, In France the unit of temperature is the Celsius

(°C) while in the United States it is the Fahrenheit (°F) and in science one uses the kelvin (K), the international unit of temperature. An equation is used to convert between them.

The origin of metrology as a drive for unifying units goes back to ancient history. The ancient Egyptians wanted to have a common unit for every measurement made at that time. The standard Egyptian scale of length (about 3000 BC) is the Cubit. The Royal Cubit was based on the Pharaoh length of forearm to the tip of the middle finger and every builder and user must have copied the Cubit for his use. Found cubit rods were about 52.5 centimeters in length. Every month, at full moon, the owners of the copied Cubits had to bring their scale for comparison with the original reference Cubit. Not doing so resulted with death punishment. Today we would say that the Cubit recalibration interval was 29 days (the length of a moon month). Some ancient metrologists in different counties just used their arm and carried this instrument wherever they went to work. Each ancient culture kept its own system of units that included length, area, volume, and weight and had laws against fraud and impartiality. Sometimes they included in the system of units the unit of time by counting days and using sun dials and some other clocks (eg. water clocks, sand clocks) which could serve in floods and agricultural predictions, or define the time given to a speaker.

Even today the logo of justice or court includes a weighing balance. Following the ancients, to keep the units in all countries in agreement, an international convention was set up (see next chapter). One of its duties is to compare the units kept in each state and to monitor them to be as defined by the convention.

Today metrology is concerned with keeping the international units at their definitions, performing international comparisons between states of the units, protecting the public from fraud measurements (e.g., correct weighing in the grocery stores), setting up quality control for correct measurements, and providing audits for those who want to demonstrate quality in their measurements.

2.1. The Metre Convention

Modern metrology started with the establishment of the Metre Convention. This is an international treaty signed in Paris in 1875.05.20 by seventeen states (note that the international convention for writing date is in the order: year-month-day). Today there are over sixty state members. At that time, the Convention's purpose was to harmonize the international units of the length, metre (can be spelled as meter in English) and mass (kilogram). The meter and the kilogram were defined using artifacts made of steel. This was organized by the BIPM, the International Bureau of Weights and Measures, set up as part of the convention. The BIPM is in Sevres, near Paris, France. Its web is https://www.bipm.org

Today, seven physical international measurement units are used as a part of the SI, the International System of units (next clause). In addition, two angles are needed, the radian (angle in plane) and the steradian (angle in volume) to complete the system. This is all that is needed to cover all measurement units used in science and technology. Every unit can be constructed from these seven. The SI is revised as necessary by the convention state members. Revisions are normally signed by the states in meetings summoned once in four years at the CGPM, the General Conference on Weights and Measures, which is the supreme authority of the BIPM. The last change to the SI was made in 2019 (see below).

2.1.1. The SI

As said above, the SI, the international system of units, is made up of seven base units. All required units which are not a base unit, are derived units and use a combination of the base units in a predefined format. The base units were redefined last time in 2019 by fixing the values of the relevant physical constants at their best-known values at that time. The fixed values are given in table 2-1. Table 2-2 lists the base units and the constants used in their definitions.

The SI defines how to write values accompanied by units and how to write derived units . See the BIPM site for the formatting [3].

Table 2-1. The exact values of the physical quantities taken from CODATA 2017[2] and fixed in 2019 to redefine the SI.

> The ^{133}Cs hyperfine transition 9192631770 s^{-1}
>
> The Speed of Light c is exactly 299792458 ms^{-1} from ^{133}Cs transition.
>
> The Planck constant h is exactly $6.62607015 \times 10^{-34}$ joule-second (J·s).
>
> The elementary charge e is exactly $1.602176634 \times 10^{-19}$ coulomb (C).
>
> The Boltzmann constant k is exactly 1.380649×10^{-23} joule per kelvin (J·K^{-1}).
>
> The Avogadro constant N_A is exactly $6.02214076 \times 10^{23}$ reciprocal mole (mol^{-1}).
>
> The Luminous efficacy K_{cd} is exactly 683 lumens per Watt.

The CGPM resolution about the redefinition of the SI was published by the BIPM[3] to take effect in 2019.

Table 2-2. The SI base units were redefined in 2019 except for the angles which were added to SI in 1960.

Measure	Base unit Name	Symbol	Units used	Constant used
Length	meter	m	s	c
Mass	kilogram	kg	s, m	h
Time	second	s		^{133}Cs
Electric current	ampere	A	s	e
Thermodynamic temperature	kelvin	K	s, m, kg	k
Amount of substance	mole	mol		Na
Luminous intensity	candela	cd	s, m, kg	Kcd
Plane angle	radian	rad		
Solid angle	steradian	sr		

2.1.2. NMI

As part of the Metre Convention, each member state is supposed to set up and support a National Metrology Institute, NMI. An example of some of the largest such institutes are NIST in the USA, NPL in UK, PTB in Germany, NIM in China, and VNIIM in Russia. In general, the duties of the NMI are to setup laboratories for the most precise measurements in their states, to provide the state with calibration service at the highest level and to participate in what is called Key Comparisons of its kept units (see below) which serve to establish trust in the NMI capability.

2.1.3. Key comparisons between NMIs

The BIPM organizes measurement comparisons among the NMIs called Key Comparisons. These comparisons are done using the NMI's highest precision measurement reference standards (realizations of the units). The outcome of the Key Comparisons (KC) is saved in the KC database (KCDB) [4]. The database can be searched by the public for the measurements outcomes obtained in the comparison and the assigned NMIs CMC (Calibration and Measurement Capability). An NMI whose comparison results are satisfactory is recognized by the CIPM (international committee for weights and measures) Mutual Recognition Arrangement (MRA). Read more in the BIPM web site, https://www.bipm.org/

2.1.4. Dissemination of units

The units defined by the BIPM should be disseminated to the NMIs or realized at the NMIs and then disseminated to secondary labs in their economies. The latest definitions of the SI can be used (in principle) to realize the units locally (without a need for traceability to BIPM). To assure that the units are realized correctly, an intercomparison is performed.
A decision that the two laboratories are compatible is judged by the parameter En (normalized error). It is defined as the difference between the two measurement outcomes divided by the uncertainty of that difference. Figure 2-1 shows a formula for two participants.

As an example of dissemination, the international time, the UTC (Universal Coordinated Time), is computed from intercomparisons (about 20 times a day) of primary atomic clocks kept in several NMIs and a large number of atomic clocks all over the world. This intercomparison is done using GPS satellites in what is called Common Mode and two-way transmissions using other types of satellites. By this method, the difference between a common signal received from a satellite is measured at the same time by several labs. The differences are reported to the BIPM which uses them to compute UTC.

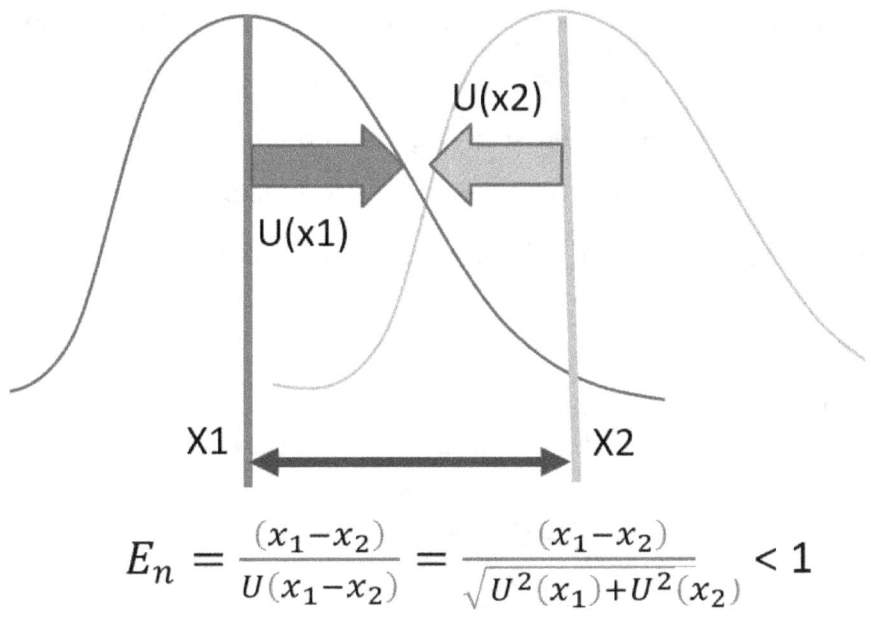

$$E_n = \frac{(x_1 - x_2)}{U(x_1 - x_2)} = \frac{(x_1 - x_2)}{\sqrt{U^2(x_1) + U^2(x_2)}} < 1$$

Figure 2-1 shows the meaning of En = $(X_1-X_2)/U(X_1-X_2)$. If the En value is in the range ±1, one cannot claim that there is a difference between the two outcomes. The rightmost term in the formula is an approximation when the expanded uncertainty can be obtained as the root sum of squares of the two expanded uncertainties. In the case of more than two participants, the En is computed vs a reference laboratory or vs a consensus value obtained from the weighted average of the participants and correlations

19

must be considered since the average value is correlated to the laboratories' values. See [11]. The uncertainty in these intercomparisons is less than 1 ns (nanosecond). The difference between the UTC and the NMIs' clocks is communicated to NMIs after a while so that it will not affect the intercomparisons. When such a cycle ends, the lab knows what its correct time was and can use it further to disseminate time and frequency to lower labs by means of calibration or signal broadcast. See Figure 2-2 below.

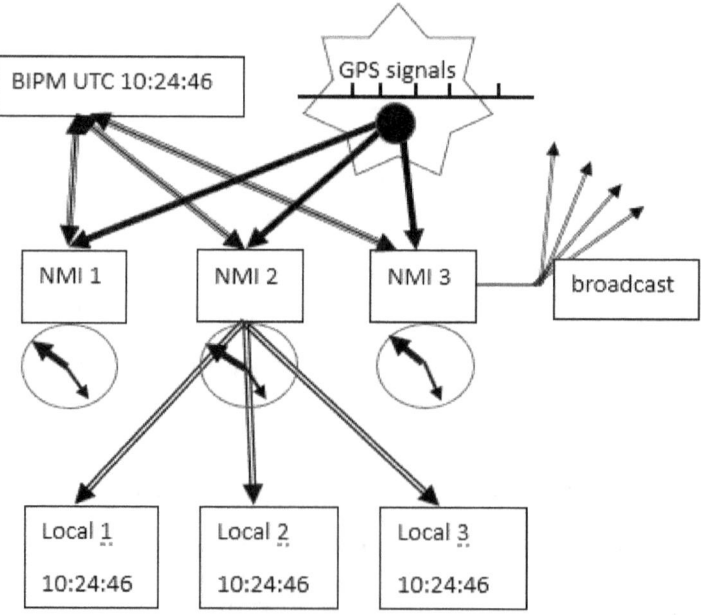

Figure 2-2. Time scale generation and dissemination. The GPS and other satellites are used to measure the time differences between the national clocks (in the figure NMI 1...3) and the GPS signals at the same time. The BIPM uses this information to generate the international time UTC (Coordinated Universal Time) and informs the NMIs about their errors. The NMIs use the errors to compute the correct local time and calibrate local clocks (in the figure: local 2) or broadcast the local time to users all over a continent (NMI 3). Sometimes the NMI clock that participates in reporting to the BIPM is a virtual time scale (not a physical clock) computed using real clocks' errors and a software algorithm [5].

2.2. Metrological international organizations

There are several international organizations related to metrology, each taking care of metrology and measurements at a different level and perspective. Figure 2-3 shows the relations between labs and the state agencies. In the figure each star represents an international organization.

Besides the BIPM which has working relations with all NMIs, the figure shows additional organizations which relate to metrology. The

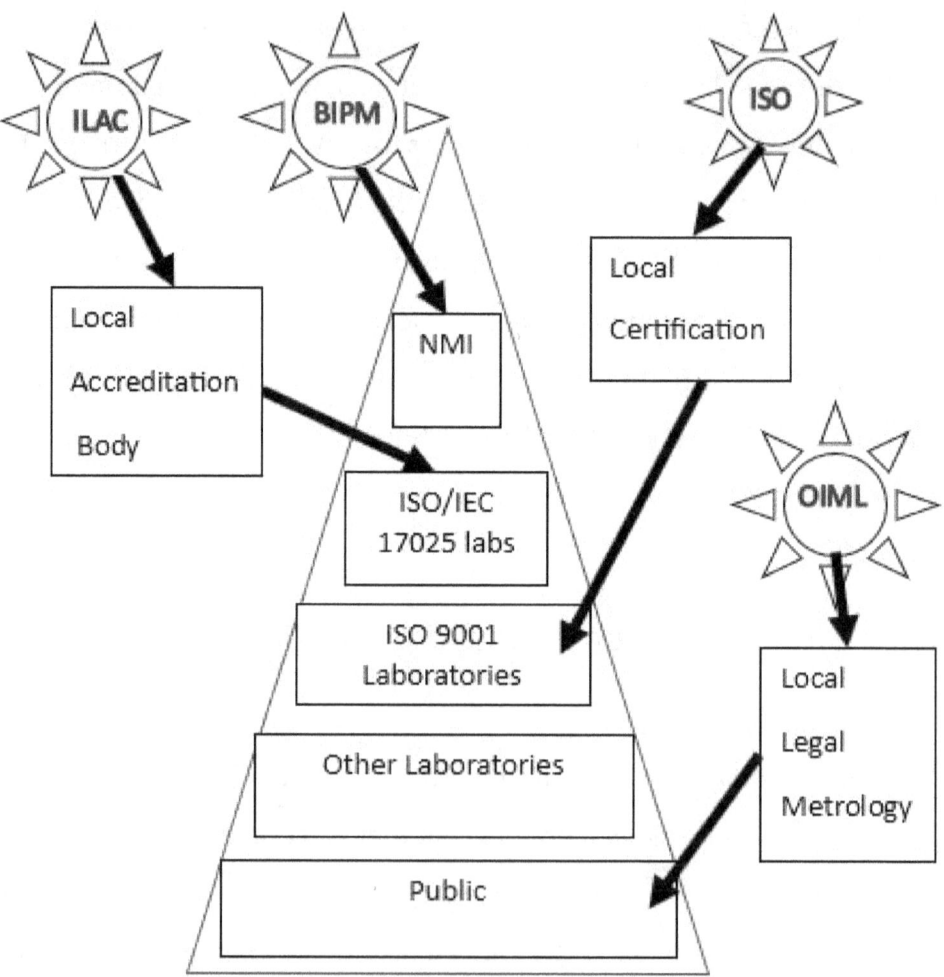

Figure 2-3. Relations between International and national metrology. The pyramid suggests that the uncertainty in calibration improves (is smaller) as one goes up the pyramid.

ILAC (International Laboratory Accreditation Cooperation) is the local Accreditation Bodies umbrella and helps international recognition of the measurements taken by any accredited lab. That is, by agreement, measurements and calibrations are accepted over state borders if done by accredited labs. The accreditation is done by auditing per the international standard ISO/IEC 17025. The accredited labs provide calibration services to the industry at their CMC (Calibration and Measurement Capability) at a level permitted by the accreditation body. Read more at https://ilac.org/.

OIML is the International Organization of Legal Metrology. Its purpose is to set up rules of how to do commercial measurement correctly, without fraud, and impartially. It is expected that each state has an agency of local legal metrology in the local government. This local agency follows and adopts the International Recommendations and other official documents published by OIML. For example, the International Recommendation R 111 instructs how to construct legal weights for different uses. For example, what weights should be used in evaluating the precision of a specific balance in a grocery. The local agency prepares a legal document based on R111 to protect the public in that state. Read more at https://www.oiml.org/.

ISO is the International Organization for Standardization. The International Standard, ISO 9001, includes instructions of how to maintain a quality measurement system in industry and similar facilities. The connection to metrology is the requirement that all measurements should be done using instruments calibrated by an accredited laboratory per ISO/IEC 17025 (see ILAC clause above). ISO nominates local bodies as their representatives to do the audits which check the compliance to the pertinent standard. Success in such audits leads to certification of the laboratory, as opposed to accreditation.

Read more at https://www.iso.org/home.html.

2.3.   Basic terms

In the following sections I list several intuitive definitions of basic terms to support preliminary discussion. For full definition see the document JCGM_200_2012.pdf in https://www.bipm.org/

2.3.1. What is a measurement?

To answer this question let us start with an example. Suppose you want to *know* what the temperature in the room is. You take a *thermometer* to measure the air temperature and you watch the displayed value which is changing from the moment you carried it into the room (because the temperature of the thermometer changes to that of the room). So, the first element in the measurement was your intension to get information about the temperature. In other words, measurement is the collection of information about a physical (or other) parameter, usually using instruments. This parameter is called the *measurand*. Normally we are interested in a numerical value, such as reading the temperature from its display as being e.g., 40 °C.

The second element in measuring was to use a trusted instrument that will do the job (*thermometer*) instead of our skin. The reason to use an instrument is that its maker constructed it in such a way that the instrument will display the temperature's value and not something else instead. Also, we hope that it will display the same temperature every time the room temperature is the same. Is our assumption correct? Could the actual temperature be different from the thermometer's reading? What is the probability or chance that the correct value differs from the read value by less than 10%? by less than 20%?

The aim of the instrument's builder is to make it in such a way that it will show the same reading when the measurand is the same. Experience shows that this is achieved only approximately, and we should trust instruments only to that degree. This can be seen by reading two thermometers from the same manufacturer at the same spot in the room. The difference between the reading (R) and the true measurand's value (M) is called the measurement error (MR). MR=R-M.

We shall come back to this issue when addressing uncertainty and measurement error later.

A critical point to consider is whether the room temperature is stable and constant or changes over time in a predictable fashion or is unstable and fluctuates unpredictably. In the above examples we assumed that the room temperature is constant and therefore saying what is the temperature, is meaningful because it is fixed overtime. But this is not the situation in most measurements. In our example, an air conditioner in the room could cool on and off and undulate the temperature. If the thermometer is fast, it can catch the momentary room temperature and if it is slow, it can only measure the average temperature. We shall address this point later.

2.3.2. What is measurement error?

The measurement error is the difference between the measured value (or generated value) and the correct value. A generated value could be the generated temperature in an oven. A measured value could be a thermometer's reading of the temperature inside the oven. That is, the measured value (or generated) minus the correct value. Since the correct value is not known with absolute precision (same as with the measured or generated values), the (measurement) error is obtained by estimation and includes uncertainty. See Figure 2-4 for a graphical presentation of the measurement error. The measured value is obtained from equation 3-1. See the discussion in chapter 3. The word 'error' suggests that there is an error, however, if one knows its value, it can be used to compute the 'true' value. A more neutral term could be measurement deviation.

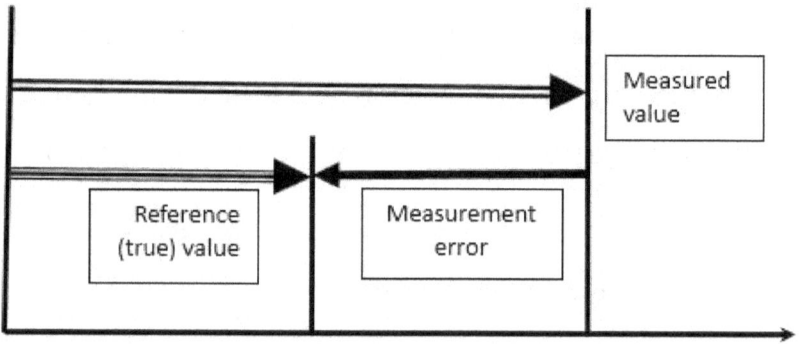

Figure 2.4. A graphical presentation of the measurement error. All values have uncertainties.

2.3.3. What is measurement uncertainty?

Uncertainty is a range about the measured value in which there is a good chance to have the correct value (remember, all readings are imprecise). This definition assumes that there is always uncertainty in the measured value (or generated value). The uncertainty must be cited together with the measured value, else, our knowledge about the measurand is not complete. For example, we should say: the temperature in the room is 40 °C ± 2 °C. This means that there is a good chance that the actual temperature is in the range (38 to 42) °C. Note the format of writing value and its unit (a space between them) and possible format for a range. We must specify what we mean by good chance and how the range of the measured value is defined. Later we shall describe two methods for the estimation of uncertainty in measurement. See a more elaborate discussion in chapter 3.

2.3.4. Precision and bias

These terms are qualitative and find use in some industries. Precision has a general meaning of how close to each other are repeated or reproduced measurement outcomes. Bias has the general meaning of how far away the average of the repeated outcomes is away from the true value.

2.3.5. What is a test?

In our discussions, a test is a measurement. A test also has more meanings, such as a check of some assumptions or conditions. We shall use in this book the word test as meaning measurement when it will not be confusing.

2.3.6. What is calibration?

Calibration is the method (or procedure) for knowing the uncertainty of a measuring instrument (or generating instrument) and its measurement error. This is done normally by comparing it to another instrument called a reference instrument, or master instrument, or a standard instrument. If one calibrates a property of a material, the reference may be a reference material and with quality uncertainty: Certified Reference Material. Usually, calibration includes the comparison of the reading by the UUT (unit under test) instrument versus the reading by a reference instrument (or a parameter compared to that of a reference material) whose precision and error at the time of comparison are known. Another possibility is to measure the output of the reference instrument and the output of the calibrated instrument and compare them. We shall discuss later what we mean by knowing the reference instrument at the time we use it in the calibration process.

Figure 2-5 shows these relations graphically. The UUT (Unit Under Test, sometimes DUT, Device Under Test, sometimes MTE, Measurement and Test Equipment) or test instrument is the instrument that is to be calibrated. A 'measure' situation occurs when the two instruments are measuring the same measurand and their outcomes are compared. A 'generate' situation occurs when one measuring instrument is measuring two generators under comparison.

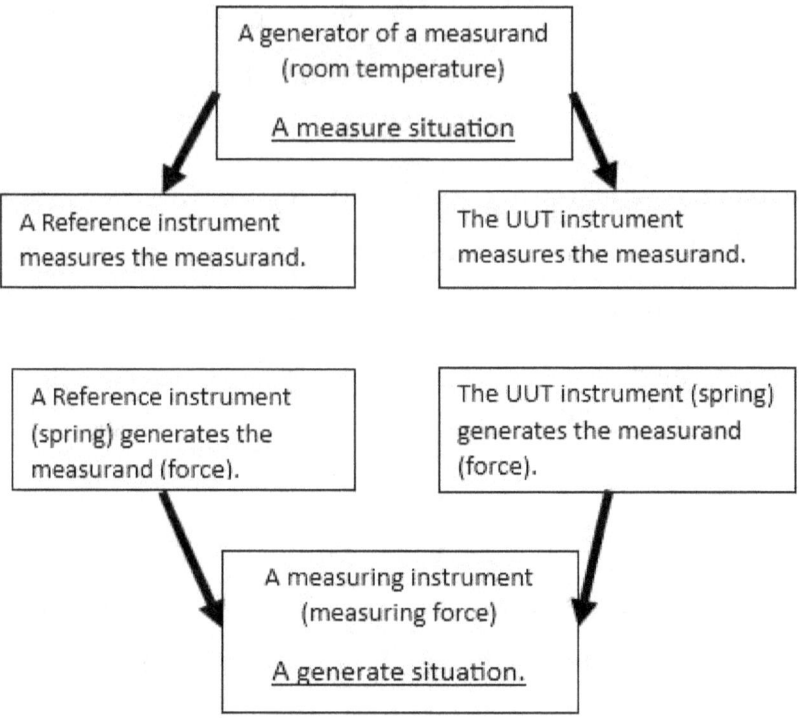

Figure 2-5. The two situations in calibration, 'measure' and 'generate'.

Measure example happens when a reference and a test thermometer measure the temperature at the same point in the room. A 'generate' situation occurs when two instruments are intended to generate the same measurand and the same measuring instrument is measuring their outputs. For example, a reference spring and a test spring are supposed to generate the same force when compressed to same amount. The two springs are measured by a one force measuring instrument (for example a strain gauge).

2.3.7. What is traceability?

In the explanation of what calibration is, I mentioned that additional instrument, called reference instrument is needed. Its uncertainty and measurement error must be known so that it can be used in calibration and help in characterizing he UUT. This reference must be more precise than the UUT, so it contributes a little to the UUT's uncertainty.

This reference instrument was also calibrated using another reference which was even more precise. Such a chain makes up a traceability chain and a metrological pyramid of less and less precise instruments when going down the pyramid. Figure 2-6 describes the situation. Each instrument is a reference instrument for the next step in the traceability chain and a test instrument for the step above it. The traceability of an instrument must be documented, that is, its calibration certificate must be available and contain all required information.

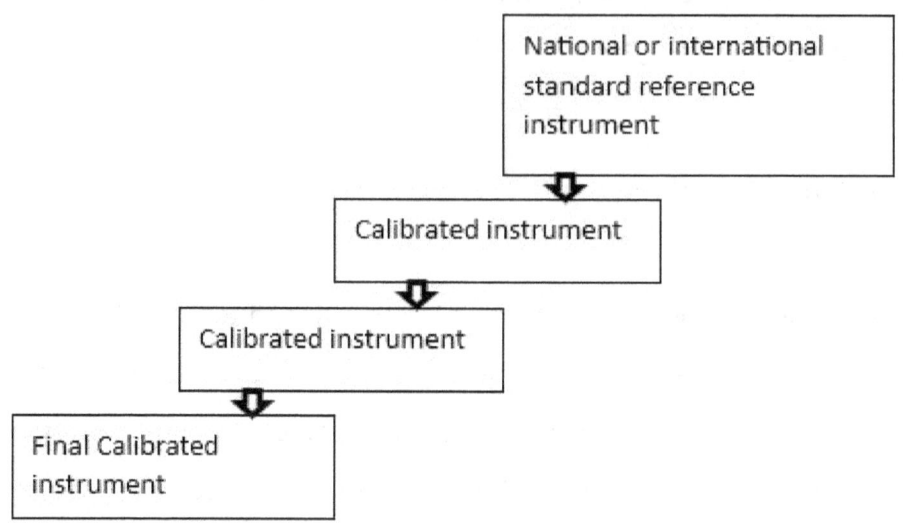

Figure 2-6. Traceability and metrological hierarchy from top national or international reference standard down to our calibrated instrument.

2.3.8. Quality assurance in metrology

To keep measurements and calibration at high quality, a quality assurance system must be actively maintained. The ISO/IEC 17025 standard provides a framework of quality assurance in testing and calibration. A laboratory that wants to demonstrate its high quality in its metrological activities can do so by requesting accreditation per this standard from the local accreditation body. In the following I cite several requirements listed in the standard for supporting quality in calibration and testing. Part of the quality assurance is to set up proper methods for the calculation of uncertainty and measurement error. See the standard for more elaborate information. The standard methods of the calculations are described in the next chapters.

2.3.8.1.   Validation of procedures and software

The laboratory must check periodically that its calibration and measurement procedures are working properly. The same must be done with the software parts developed by the laboratory and which are not widely used by other laboratories (not off-the-shelf software).

2.3.8.2.   Proficiency testing

Under this item are listed several methods that a lab can use to prove the capability of the lab and its staff. For example, the lab must perform periodical internal audits to check if its calibration procedures are performed as intended by the procedure authors. The written procedure must be reviewed periodically by the lab to check that this is still the situation.

2.3.8.3.   Intercomparisons

To keep confidence in the calibrations and test, the laboratory should compare its measurement outcomes with other laboratories. Several possibilities for such intercomparisons exist and are listed below. See Figure 2-2 for a discussion on time comparison.

A laboratory can compare the calibration results, as stated in their calibration certificate they issued with the calibration certificate of another laboratory that measured same measurand at about the same time (time is important because of drifts).

A laboratory can participate in an intercomparison organized by an organization accredited for such intercomparison activity.

A laboratory can participate in a bilateral comparison, for example between its' outcomes and the outcomes of a recognized reference laboratory such as an NMI.

2.3.8.4.   Checks before and after the use of an instrument.

When a laboratory is calibrating on the customer's premises, it is recommended that it checks its instruments before and after the service so if there is a large measurement error it is easier to find the reason.

2.3.8.5.   Intermediate calibrations

The instruments used by the lab are drifting and changing over time. To check that such an instrument is still performing satisfactorily, it is assessed in between its officially planned calibrations. Such calibrations are called intermediate checks. Normally only a small part of the instrument's functions is checked, a part that is sufficient to indicate that the instrument is   still   precise   as   needed   to   perform   its   functions.

3.   Estimating uncertainty

As defined in clause 2.3.3., uncertainty is a range about the measured value in which there is a good chance to have the correct value. Now we need to be more precise in our definitions to be able to estimate uncertainty. The starting point is the probability density function of the outcome (measured quantity or any uncertain quantity).

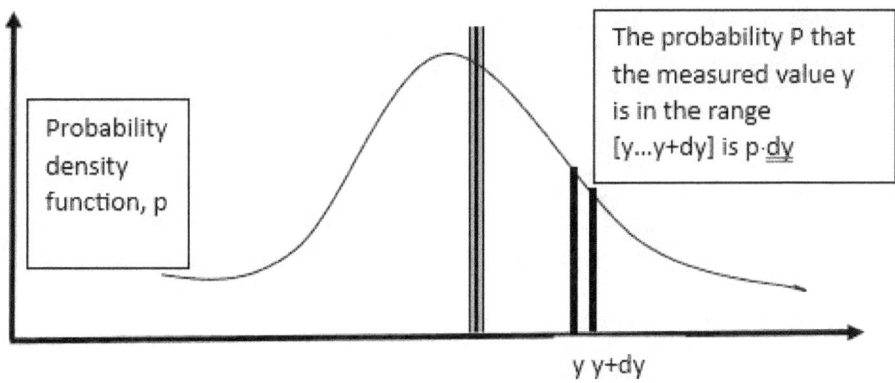

Figure 3-1. The measurement probability density function.

Figure 3-1 shows an example of the probability density function of the possible outcomes of a specific measurement. Suppose we repeat the measurements and draw a graph of the probability density function. This function shows the probability of each outcome.

The probability, P, for an outcome to be at point y is P = p · dy, where p is the probability density function and y a possible measured outcome.

The area under the probability density function is 1 (coverage probability) since the total probability to find the measurement somewhere is unity.

The average (or expectation) of y is defined as the weighted average outcome of the measurements. The weighted average of y, μ, is given by:

$$\mu = \int_{\min y}^{\max y} p \cdot y \, dy \qquad\qquad (3\text{-}1)$$

and it points to the center of gravity of the function p. Note that the average does not have to be at the pick of the graph as shown in figure 3-1. The maxy and miny limits are the endpoints of the possible outcomes.

Another important parameter is the standard deviation (sd) of this density function.

$$sd^2 = \int_{\min y}^{\max y} p \cdot (y - \mu)^2 \, dy \qquad\qquad (3\text{-}2)$$

One type of uncertainty defines uncertainty as sd. This type of uncertainty is called Standard Uncertainty and is usually denoted by small u. Therefore, when reporting measurements using standard uncertainty, one may say that the measured value or outcome is:

$$\text{Outcome} = \mu \pm sd = \mu \pm u \qquad\qquad (3\text{-}3)$$

That is, one may say that the outcome is the average plus/minus the standard deviation of the density function. Note that if the density function is normal (or gaussian having the form of a bell, see Figure 3-4), the sd's coverage probability is 68.27% , that is, there is a chance of 68.27% that the outcome is within the range ±sd about the average (which is 2sd). The coverage probability of the standard uncertainty, u, is unknown if the density function is not given.

The other type of uncertainty is based on specifying the coverage probability, normally 95% for calibrations and tests. It is called Expanded Uncertainty and there are two ways to define its value. Figure 3-2 shows the two definitions. When there is only one pick the two definitions result in the same expanded uncertainty.

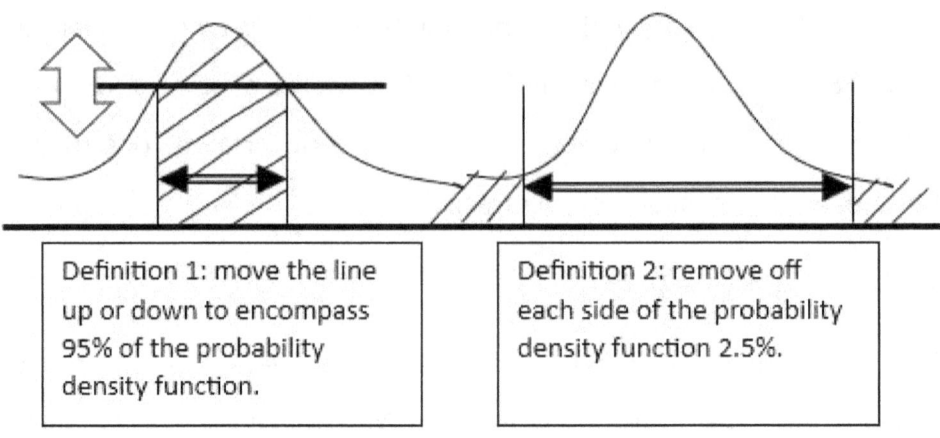

| Definition 1: move the line up or down to encompass 95% of the probability density function. | Definition 2: remove off each side of the probability density function 2.5%. |

Figure 3-2. Definitions of the expanded uncertainty.

Note that if the probability density function is not symmetrical, the positive and negative uncertainties are different because the average may not be at the density's function peak. The standard uncertainty, on the other hand, is always symmetrical by definition.

ISO Guide to the Expression of uncertainty in measurements (GUM) [6] describes a method of estimating uncertainty. Other methods are described in [7] and [8]. The first step in all methods is to write the measurement equation. This equation describes how the outcome, y, is obtained from all the input quantities (influence factors), x_i, affecting it:

$$y = f(x_1 .... x_N)$$ (3-4)

For example, suppose we want to measure the temperature of the coffee inside our cup. We expect that the reading of the temperature depends on the uncertainty and the error of our thermometer (x_1), on the time elapsed since the thermometer was calibrated (it may drift) (x_2), on the length of time we measure (it takes time for the thermometer to be at the coffee's temperature) (x_3), on the position of the thermometer in the cup (close to surface it cools faster) (x_4). You are welcome to propose more influences and input quantities. In principle we could construct a complicated measuring equation taking all this into consideration. Such an equation could be an approximation to the real world and thus contributing to the measurement uncertainty by being only approximate.

I shall list below some input and influence quantities that should normally be considered when estimating y and its uncertainty:

- Repeatability (variation in repeated measurements),
- Reproducibility (variation when the setup and probably the operator are changed),
- Human factor (how much the operator affects the measurement),

- Error and uncertainty of the reference instruments (or material) revealed when calibrated,
- Aging of the reference instruments (change of their error and uncertainty overtime),
- Resolution in the UUT reading, as read from display or as read by a computer (or the precision of the setpoint of a generator),
- Hysteresis (a measurement may depend on the values of past measurements and show different readings when the earlier measurement was higher or lower than the final nominal),
- Environmental conditions (for example temperature, humidity, barometric pressure, power supply, vibrations, air flow, statics.),
- Limited calculation capability (if a calculator has 10 digits you will be able to calculate uncertainty at the 5-digit position because when calculating sd, it is required to raise to power of 2 and then taking square root when estimating uncertainty, see equation 3-5. When using 16 digits the uncertainty is calculated at the 8-digit position),
- Imprecise measuring procedure resulting in uncertainty and error.
- Not measuring the correct measurand.

44

3.1. Estimating standard uncertainty per GUM

Following GUM [6], to estimate the standard uncertainty, we need to know the standard uncertainties of the input and influence quantities in the measurement equation (equation 3-4). Following GUM, there is no need to know the quantities' errors and GUM assumes that they are all zero. Also, there is no need to know the shape of the probability density function. From first order Tailor expansion of y, equation 3-4, and the laws of combining standard deviations, we may deduce that the standard uncertainty of y, u(y), is obtained from the standard uncertainties of the x_i input quantities and is given by the equation:

$$u^2(y) = \sum_i^N \left(\frac{\partial f}{\partial x_i}\right)^2 u(i)^2 = \sum_i^N (c_i)^2 u(x_i)^2 \qquad (3\text{-}5)$$

The c_i, which are the derivatives, are called the c sensitivities. They reflect the amount of the influence of each input quantity uncertainty, $u(x_i)$ on the final (combined) standard uncertainty. This equation assumes that the x_i quantities are statistically independent and uncorrelated.

Suppose that an instrument is affected by the ambient temperature, and by the relative humidity, and by the barometric pressure. These quantities are correlated since they affect each other. For example, the temperature may change if the barometric pressure changes. It is easy to test if there is a correlation between two quantities, if one of them can be changed over a range of values and the other is monitored. Figure 3-3 show how a correlation between two quantities looks graphically.

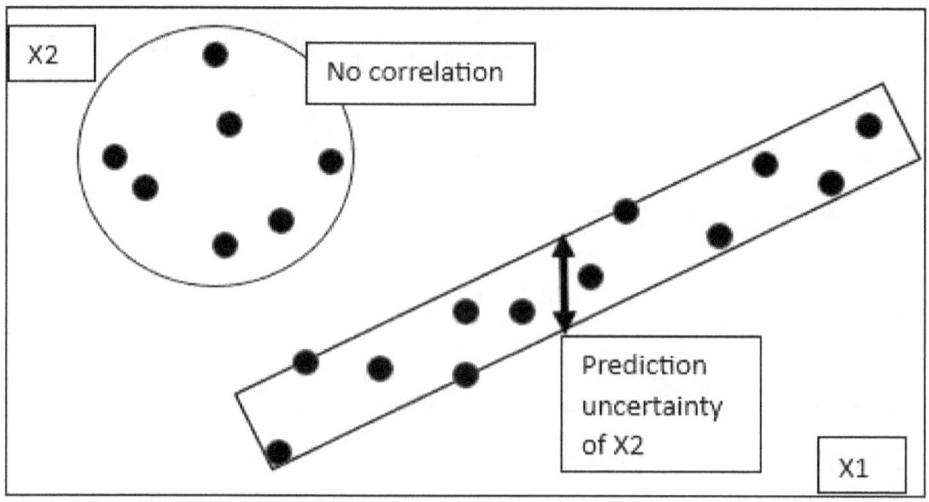

Figure 3-3 shows the values of X2 when X1 is changed (black points). The values fit in a rectangle showing a positive linear correlation (the rectangle's slope is positive). It is possible to predict the value of X2 to some extent from the values of X1. The prediction uncertainty range is the vertical width of the rectangle as shown. If the prediction is perfect, one obtains a straight line. If the points fit in a circle as shown, there is no correlation, and one quantity cannot be predicted from the other better than a guess.

Note that Input quantities can be predictable (one quantity can predict the other to some extent) even if their linear correlation coefficient is zero. For example, when X1 and X2 form a circumference of a circle. The linear correlation coefficient is zero, but it is possible to predict X2 from X1 because one is the exact function of the other.

When the input quantities x_i are linearly correlated, the computation of $u(y)$ must consider the correlation coefficients r_{ij} between the i and j input quantities:

$$u^2(y) = 2 \sum_i^{N-1} \sum_{j=i+1}^{N} \frac{\partial y}{\partial x_i} \frac{\partial y}{\partial x_j} u(x_i)u(x_j)r(x_i, x_j) \qquad (3\text{-}6)$$

The linear correlation coefficients $r(x_i, x_j)$ are defined as:

$$-1 \le r(x_i, x_j) = \frac{\sum_{i,j}(x_i - \mathrm{avgi})(x_j - \mathrm{avgj})}{\sqrt{\sum_i(x_i - \mathrm{avgi})^2 \sum_i(x_j - \mathrm{avgj})^2}} \le +1 \qquad (3\text{-}7)$$

47

Where avgi is the average value of x_i etc. The range of values of r is [-1...+1]. When r is -1 the two quantities act to cancel each other, and the uncertainty is better (smaller). One should always try to construct a situation where r = -1 to reduce uncertainty. For example, when we measure the length of a rod with the same bad gauge (large measurement error) and look at the difference of outcomes from measuring the positions of both sides of the rod, the readings should cancel each other, and an improved measurement can result as opposed to a measurement with two gauges.

When all r_{ij} = 1, equation (3-6) is simplified and the outcome standard uncertainty becomes the sum of the input quantities' uncertainties, that is:

$$u\,(y) = \sum_i^N c(x_i)\,u(x_i) \; . \hspace{4cm} (3\text{-}8)$$

Simplified cases occur when the measurement function in equation (3-4) is the sum of the components X_i or their multiplication. In the case of sum, the derivatives are 1 and equation (3-5) simplifies to equation (3-9) below:

$$u^2(y) = \sum_i^N u(x_i)^2 \hspace{4cm} (3\text{-}9)$$

In the case of multiplication (and division), it becomes the root sum of squares of the relative standard uncertainties of the input quantities as given below:

$$\left(\frac{u(y)}{y}\right)^2 = \sum_i^N (u(x_i)/x_i)^2 \qquad\qquad (3\text{-}10)$$

Which is the sum of squares of the relative standard uncertainties.

3.1.1. Assumptions about probability distributions

To compute the standard uncertainty of the output, $u(y)$, using equations (3-5) or (3-6) we need to know the standard uncertainties of the components, that is, $u(X_i.)$

Table 3-1. Several distributions and their standard deviations.

Distribution	Example of Use	parameter	Standard Deviation
Rectangular, constant	Resolution of digital display. Also used,	a = the range of least	$\dfrac{a}{\sqrt{12}}$

probability over the range a	when in absolute ignorance.	significant digit	
Triangle	Control about the top of the triangle.	b = width of half triangle's base	$\dfrac{b}{\sqrt{6}}$
Gaussian	Calibration certificates. a distribution combined of many other distributions	σ parameter defining the distribution	1σ
N Repeated measurements	Sample Standard Deviation of the Mean, sdm	Avg $= \sum x_i \ /N$	sdm^2 $= \dfrac{\sum_1^N (x_i - avg)^2}{N(N-1)}$
Student with n>2 degrees of freedom	Combined standard uncertainty assumed by GUM	n=degrees of freedom	$sd = \sqrt{\dfrac{n}{n-2}}$

U-Sine	Used in RF reflection, may be obtained from rectangular distribution of the angle (phase)	One cycle width	$onecycle/\sqrt{2}$
Two picks	May be used in flip-flop distribution	Half distance between two picks	Half distance between two picks.

To this end, we must know the probability distributions of the components and then use equation (3-2) to find the standard deviation of the distribution. In most cases it is too complicated for a calibration laboratory to figure out the probability distribution function and plausible assumptions are made as exemplified in Table 3-1. Table 3-1 lists seven examples of commonly used distributions. More distributions can be found in [9].

The following Figure (3-4) shows the graphical appearance of the normal and student distributions.

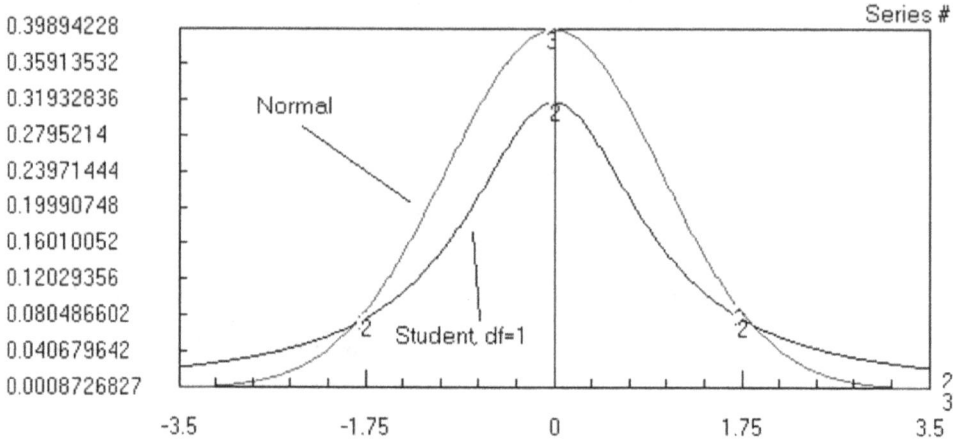

Figure 3-4. A graph of normal and student distributions with 1 degree of freedom. Student distribution with higher degrees of freedom lies between these two distributions. When the degrees of freedom approach infinity the student distribution approaches the normal distribution. This graph was drawn using MetroVal program [11].

GUM assumes that the standard deviation of the outcome y is student distribution. This is an approximate assumption and has uncertainty. For example, if the dominant contribution X_i has a rectangular distribution because it reflects a digital display with low resolution, GUM assumption is a bad approximation and is in error. Back to GUM. To define the specific student distribution of the outcome, one must define a parameter called 'degrees of freedom'. It can be a real number (not just an integer as with the distribution due to repeated measurements), and therefore is called the effective degrees of freedom. It serves as a shape parameter of the distribution. To estimate the effective degrees of freedom, GUM recommends using the Welch-Satterthwaite formula:

$$\nu_{eff} = \frac{u^4}{\sum_i^N \frac{u_i^4}{\nu_i}} \qquad\qquad (3\text{-}11)$$

Where ν_{eff} is the outcome effective degrees of freedom.
ν_i are the degrees of freedom of the standard uncertainty u_i. u is the outcome standard uncertainty. It is N-1 for N repeated measurements and infinity for most other distributions.
The ν_{eff} will be useful later when computing the expanded uncertainty per GUM method.

This formula (3-11) is an approximation for the case of independent input quantities. GUM does not provide an equation for the case of correlated quantities.

In practice one assumes that v_i is infinite except for the components of repeated measurements where it is N-1. This means that in the denominator in equation (3-11) one uses only the repeated measurements components since all the others are 0 (= 1/infinity).

3.1.2. Assumptions about c sensitivities

To compute the standard uncertainty per GUM (see Equation 3-5), one must know the c sensitivities. For example, if one wants to estimate the influence of ambient temperature on a dynamometer (force measuring instrument), he could do the test in an environmental chamber and change the temperature while reading the dynamometer's display. This may be a complicated procedure and therefore the results should be provided by the manufacturer as part of the dynamometer's specifications. Many laboratories skip this step and assume that c=1 (if there is not involved a unit conversion). Unit conversions are needed when the u(y) and $u(x_i)$ are given in different units and the c_i are used to convert the $u(x_i)$ units to the u(y) units. For example, in the case of only one input quantity we have u(y)

$= c_1u(x_1)$. Suppose we measure temperature, and we want $u(y)$ to be in kelvin, but the thermometer reads in Fahrenheit, and so we have $u(x_1)$ in Fahrenheit. Then a change of 1 °F is equivalent to a change of 5/9 K and c_1 = 5/9 to balance the equation. We are interested only in changes since the c_i are derivatives. The result is $u(y) = (5/9)*u(x_1)$

There are cases for which there exists a published standard procedure (by a well-recognized standard authority) for a test or calibration and the estimation for the involved uncertainties, sensitivities, and errors. in such a case I recommend following the standard even if it adds uncertainty. The reason is that if everyone uses the standard it is easier to compare results between different measurements. For example, see the standard DIN EN 14212 for the measurement of air pollution by Sulphur dioxide[10].

For these reasons it is recommended to use instruments that come with documented specifications for sensitivities and uncertainties for all input quantities.

3.1.3. Estimating expanded uncertainty per GUM

The GUM and ISO/IEC 17025 are asking that the expanded uncertainty is the main uncertainty published in the calibration certificates. To estimate the expanded uncertainty, one needs to know the output probability distribution. To avoid the assessment of this function, GUM assumes that the outcome is always student distribution. To this end we must know the degrees of freedom of the outcome standard deviation. This was already addressed in equation (3-11). We must find the coverage factor k=U/u, where U is the expanded uncertainty and u the standard uncertainty. This is given in statistical tables for specific degrees of freedom. GUM also gives such a table to help find k for the required degrees of freedom and the coverage probability. The coverage probability for calibrations required by GUM is 95%.

Table 3-2 show the coverage factor k for several coverage probabilities and degrees of freedom. The 95% and 95.45% are both sited because some labs are using k=2 leading to 95.45% instead of 95% coverage probability in their computations. This makes a difference at low number of degrees of freedom. Note that k for 30 degrees of freedom is close to k for infinity and therefore it is sufficient to make no more than 30 repetitions when repeating measurements for 95% coverage probability uncertainty estimation. For normal distribution coverages of 68.27%, 95.45%, and 99.73% in Table 3-2 correspond to standard deviation of 1, 2, and 3 sigma.

Table 3-2. Coverage factor k for several coverage probabilities and degrees of freedom.

Degrees of freedom	k for 68.27% coverage	k for 95.0% coverage	k for 95.45% coverage	k for 99.73% coverage
1	1.84	12.71	13.97	235.8
2	1.32	4.3	4.53	19.21
3	1.2	3.18	3.31	9.22
10	1.05	2.23	2.28	3.96
30	1.02	2.04	2.09	3.27
infinity	1	1.96	2.00	3.00

3.1.4.   Summary of the GUM method

The following are the main steps in calculating the expanded uncertainty per GUM:

- Write the measurement equation (3-4),

- Supply the parameters for the standard uncertainties (3-5) or (3-6)
- Compute the output standard uncertainty u(y) and its effective degrees of freedom, equation 3-11.
- Find the coverage factor k assuming the output is of student distribution (e.g., Table 3-2),
- Find the expanded uncertainty, U, from U=k*u,
- Report the results (see clause 3.5.4). At least report U and k and declare that the uncertainty is for 95% coverage probability. Also supply the measurement error.

Figure (3-5) shows one specific way to realize the above steps.

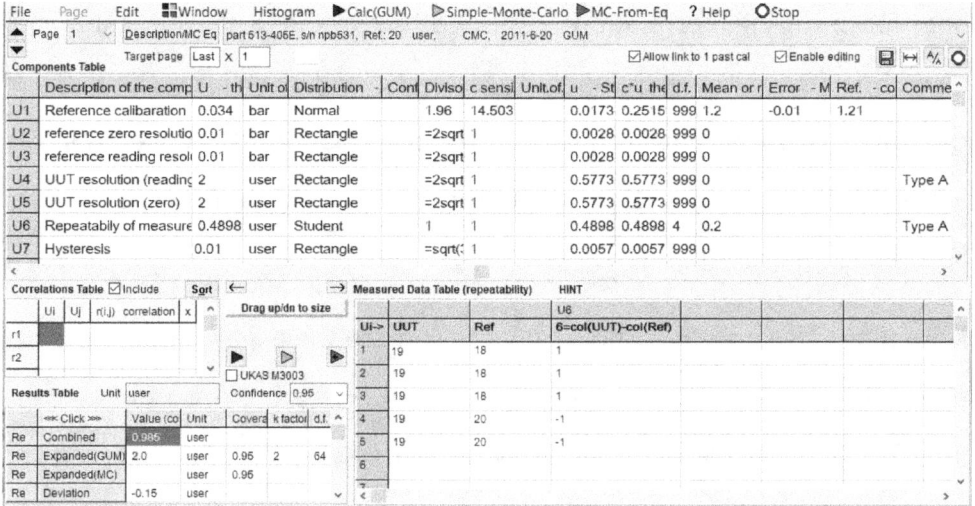

Figure 3-5. The figure shows the calculation method used by the program MetroVal [11] which follows the above steps when calculating standard and expanded uncertainties per GUM. The shown tables together are referred to in the GUM as 'the uncertainty budget'. The top table lists the components (input quantities). It shows, for each input quantity, given on a separate row, its expanded uncertainty, unit, probability distribution, sensitivity, coverage factor, degrees of freedom, and measurement error. The user or a software subroutine must supply a minimum of information (there is some redundancy) that enables the application to do the calculations. A table is provided for entering repeated measurements and

one for entering correlations. The bottom table lists the estimated standard uncertainty, expanded uncertainty, and measurement error of the real part of the output quantity. The imaginary part is covered in the figure and not shown. Data is entered manually or automatically from automatic measurements.

Note that GUM does not show how to estimate the expanded uncertainty when the input quantities are correlated or when the output density function is not student. These drawbacks are solved on the fly using the Monte-Carlo method described in the next clause.

3.2. Estimating expanded uncertainty using Monte-Carlo method

The starting point is again the measurement Equation 3-4. One must know the distribution of each input quantity. However, the standard deviations and the sensitivities are not needed and not used. The method involves generating random numbers which have the distribution of the input quantities (already including measurements errors) and have the right correlations between these quantities. These random numbers are substituted into the measurement equation and a value for y is computed.

When the computation of y (Equation 3-4) is repeated, say 10000 times, one obtains 10000 values of y which are used to prepare a histogram of y. From this histogram we can calculate the standard deviation, the expanded uncertainty, the average output quantity and use it for the evaluation of other parameters such as risk analysis of the outcome. This is addressed in clause 4.3.

The Monte-Carlo method is a simpler method from the point of view of the user because the computer does all the required calculations once the measurement equation is entered. The software must know how to generate the random numbers for the provided, possibly correlated, distributions.

As an example, let us assume that the measurement equation is $Y = 3X_1 + 5X_2 + X_3^2$, that is, Y depends on 3 input quantities, and we assume for the simplicity of demonstration that all have a rectangular distribution with average zero and widths ±1, ±2 and ±3. Suppose that we have 100000 repetitions of 3 different random numbers (one for each X_i). Suppose that in repetition 1, $x_1=1$, $x_2=1$, $x_3=1$, therefore $y(1)=9$. Suppose that the next repetition generated $x_1=1$, $x_2=2$ and $x_3=3$. Then $y(2) = 3*1+5*2+3^2 = 22$. Repeating it 100000 will provide the data for a histogram. Figure 3-6 shows this histogram. The program removed 2.5% on both tails of the distribution to obtain the range of 95% coverage probability. The estimated mean of the distribution was 3 and not zero (despite the fact that each Xi mean was zero) because of input x_3 which is squared, and the expanded uncertainty is about 12 (2-digit resolution). This is because uncertainties are reported with 2 significant digits. For example: 0.0034 m. More significant digits do not contribute to information because the uncertainty in the uncertainty may be quite large (even 30%) and additional digits are insignificant.

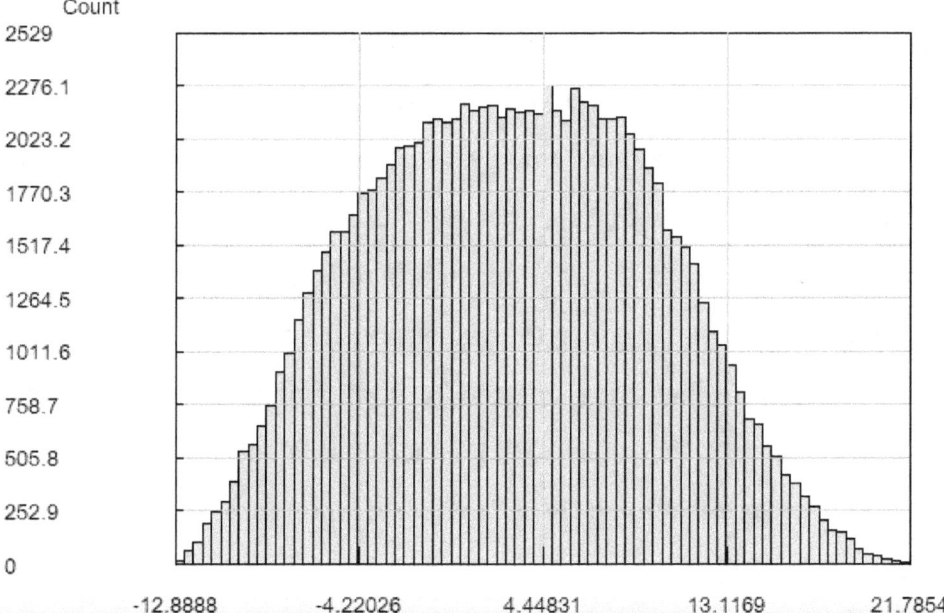

Histogram: Sample size 100000 points, 80 Slots. Min & Max values: -12.8888 & 21.7854

	>>> Click <<<	Value (computed)	Unit	Cove	k fac	d.f.	d	Low	High
Re	Sigma (MC)	6.59	user						
Re	Expanded(GUM)	12	user	0.95	1.96	1.2E+03			
Re	Expanded(MC)	12	user	0.95	1.85			12	12
Re	Mean (MC)	3.0	user					-8.9	15.4
	Imaginary-Part								
Im	Sigma (MC)								

Figure 3-6. The outcome of the Monte-Carlo simulation was obtained for the above example by the program MetroVal [11]. The high and low endpoints (-8.9 and 15.4 about a mean of 3.0) are not symmetrical as can be seen in the histogram. However, because the program displays expanded uncertainty with 2 significant digits only as required, this cannot be seen in the value of the expanded uncertainty displayed as 12. Note that every time the simulation runs, the values and the shape of the histogram may differ because they are generated from random vectors. If the change in the uncertainty is less than 10%, it is acceptable and no additional run with more cycles is needed. Most people do not care about change in uncertainty by 10%.

Note also that even with 3 rectangular input quantities the histogram approaches normal distribution. This does not happen with student distributions with 1 degree of freedom because they do not obey the central limit theorem (stating that with some conditions the distribution of the sum of many input quantities tends to approach to the normal distribution) and even with 2 degrees of freedom the limit is approaching normal distribution very slowly.

3.2.1. Correlated components

To compute the output quantity when the input quantity is obtained from correlated quantities, the random number vectors standing for the input quantities must be correlated the same way. This is carried out in the program MetroVal [11] by preparing random number vectors that have a fraction of their length the same random numbers as required by the correlation value. That is, the program replaces portions of the random vectors with common numbers. The length of the common portion of the vectors defines the correlation coefficient. See Figure 3-7.

3.2.2. Output of several outcome quantities

Here we describe the measurement of several quantities simultaneously. Such a measurement situation is described in [8] and [11], while [15] describes the algorithm to realize the calculation. The calculation can be realized by having a unique measurement equation for each outcome quantity. Therefore, the non-trivial setup is when there are common input quantities for each of several output quantities. For example:

Y1 = f(X1, X2, X5)

Y2 = g(X2, X4)

Y3 = h(X2 X3, X4, X5)

Here we have three output quantities (Y1, Y2, Y3) that use different functions (f, g, h) for five input quantities (X1, X2, X3, X4, X5). Therefore, the quantities Y1, Y2, Y3 may be correlated to some extent.

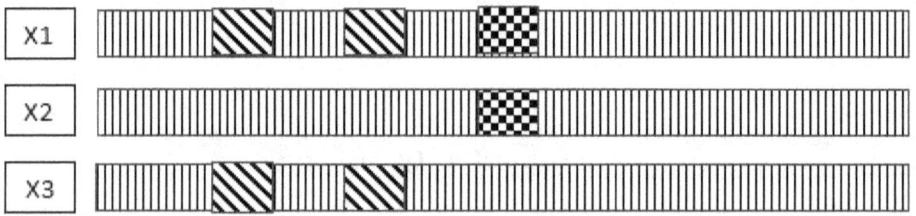

Figure 3-7. The first three vectors and portions of common random numbers are shown. Assuming that the random vectors contain 15000 numbers (to be used in 15000 calculations of each of the Y's) and the common portions are of 1000 numbers, than vector X1 is correlated with vector X3 about r = 0.133 = 2*1000/15000 ; X1 is correlated with X2 about

r = 0.066 = 1000/15000; and X2 is uncorrelated with X3 about r=0 (no common numbers).

To realize such a measurement, the program MetroVal [11] prepares a random vector (see clause 3.4.1) for each of the Xi and uses the Monte-Carlo method for each of the Yi. That is, in the above example, if the 3rd random number for X2 is 7 it will be substituted in X2 in all three Yi in the 3rd cycle of calculations[12]. Figure 3-7 describes such a situation.

3.3.   General issues concerning calibration

Here we introduce some general issues related to calibration.

3.3.1. How many points and which points to calibrate?

We are interested in calibrating as few points as is sufficient. To answer the question about the minimum number of points, one should know the extent of non-linearity of the process to be measured. Figure 3-8 is used to describe the point. Suppose we decide to calibrate at two points (black circles in the figure). Two points will result in a straight-line calibration curve (black line). The deviation between the straight-line and the actual curve hints if we need an additional point in the middle of the

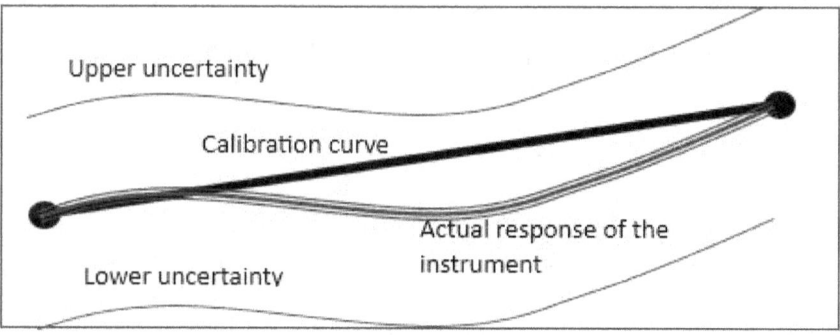

Figure 3-8. How to decide on the number of calibration points. See text.

calibration range. Also, the risk that the straight-line lies outside the upper or lower uncertainty is a factor in the decision. In the example of Figure 3-7 this is not a major problem unless it also violates the tolerance limits (not shown). Information about nonlinearity is sometimes supplied by the manufacturer or obtained from prior experience. In some cases, for example, with the calibration of Platinum Resistance Thermometers (PRT) the calibration lab is expected to supply a complicated calibration polynomial.

3.3.2. Autocorrelation and time between repeated measurements

Autocorrelations means that a measurement process is correlated with itself. Usually, by this we mean that a measurement outcome depends on earlier measurement. This may happen for two reasons. Either the instrument remembers the earlier measurement (as in the case of instrument hysteresis) or the measurand cannot change abruptly when going to a different point.

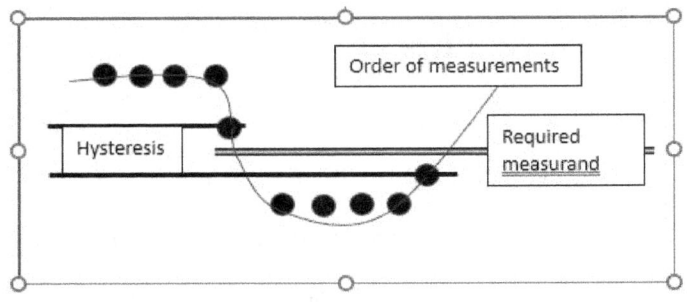

Figure 3-9. Order of measurement in case of defining instrument hysteresis. The black circles are the instrument's outcomes and are different from the required measurement. See text.

If the instrument is affected by earlier measurements, we should try to measure several measurements above the value and then take the required one and next, several measurements below the value and then the required measurement. This is a general practice also to estimate the range of hysteresis. This is described in Figure 3-9.

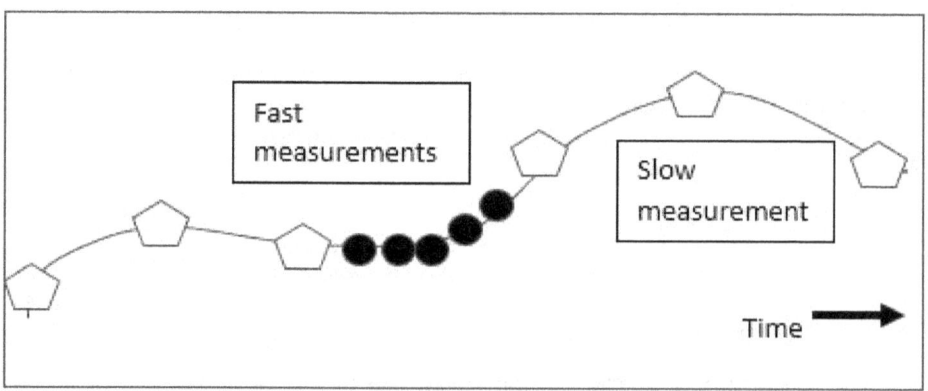

Figure 3-10. Many fast measurements in a short time interval may not always correctly describe an unstable process. See text.

If the process changes over time, we should wait enough time between measurements (Figure 3-10).

3.3.3. CMC

Accreditation bodies publish the CMC values (Calibration and measurement capability) of their accredited laboratories based on the laboratory's capabilities. These are values assigned administratively by the accreditation bodies. The laboratories can publish, in calibration certificates, only the higher of the estimated expanded uncertainties and CMC values for each calibration point. Normally they use the CMC in their publications.

3.3.4. Reporting Calibration results

The metrological and quality assurance requirements of presenting measurement or calibration information in calibration certificate (or calibration report) are given in ISO/IEC 17025. The metrological aspects that should be reported in a calibration certificate are:

71

- Identification of the measurand and the operator,

- Date of calibration (because of drifts).

- Method of measurement (setup, interval between measurement, duration of a measurement, and the number of repetitions if affecting outcome),

- Traceability of the reference instruments or reference materials. Normally, data about the calibration of the reference instruments (this information is sometimes saved with the lab and is not published in the issued certificate but may be provided upon request),

- Measurement errors

- Measurement uncertainties (at least expanded uncertainty with positive and negative values if not symmetric, k coverage factor), coverage probability (95%), and sometimes degrees of freedom when meaningful. Expanded uncertainty published with 2 significant digits and the errors have the same position of last significant digit as the uncertainty.

Other information may include the method for assessment of compliance with specification and a decision on compliance (chapter 4), and the date of next calibration based on analysis of drifts (chapter 5) or based on authority requirement.

3.3.5. Adjustment of an instrument

In some cases, the user of an instrument may want to have its readings reflect the true value with minimized measurement error. To this end, he may adjust the instrument's reading to fall within the required range. Normally, after such a readjustment of the instrument it should be followed by recalibration. Permission for adjustment when justified and following recalibration must be agreed beforehand, normally in the contract review between the calibration laboratory and the user.

4.   compliance with specification

Here I discuss compliance with specification. Compliance methods should be addressed in the Contract Review (Required by ISO/IEC 17025) that a calibration laboratory is supposed to sign with its customers. The standard does not require declaring compliance with specification in the calibration certificate, but how to state it if included in the calibration certificate (e.g., by user request).

4.1. Specification (tolerance, limits, max permitted error)

To describe the instrument's performance, the manufacturer supplies tolerance or limits specifications about the extent of possible errors. A government agency may define maximum permitted error in critical applications. If so provided, the tolerance (or limits) may be considered as a contract with the manufacturer or the testing laboratory. Instrument tolerances for a specific project can be defined by the user of the instrument and may be different from the manufacturer's specifications. In this case the specifications should be listed in the contract review (obligatory for accredited labs) and given in the calibration certificate.

Specification can be given as a fixed region or a confidence interval. If given as a fixed region, the request is that 100% of the outcomes of the measurements are within this region. If less than 100% (e.g., 95%) is specified, then some of the measurements may be out of tolerance (not complying with specification). The specifications may refer to one- or two-sided limits or tolerances. For example, one may specify that the room temperature in the laboratory should be within the tolerance (23±3) °C or within the limits (20 to 26) °C. Note the way to write down the units of a range and one space between the value and the unit. Also, tolerance normally refers to the deviation from a reference value while limit refers to a maximum (or minimum) permissible value.

4.2. Compatibility with specifications

In many cases the owner of a measurement instrument wants to know if the instrument outputs measured values within a specified range about the correct value. Note that this range is not uncertainty and may be in contradiction with uncertainty. Such a requirement arises, for example, in manufacturing when a technician needs to know if a manufactured item meets specification. A failure may result in readjustment of the production process and in adjustment and recalibration of the involved instruments (see Clause 3.3.5). This procedure is referred to as process control. We shall not deal with this subject.

4.3.   Conformity assessment

The document ILAC G8 [13] describes several methods of assessing conformity with specifications. But the basic principle is that any method agreed upon by the parties (calibrating lab and owner/user) is acceptable. The document refers to "Guard Band" methods and to "Risk Assessment" methods.

4.4.   Conformity method using risk assessment.

By 'risk' we mean that there is a possibility that an incorrect acceptance or rejection of the outcome has been made.

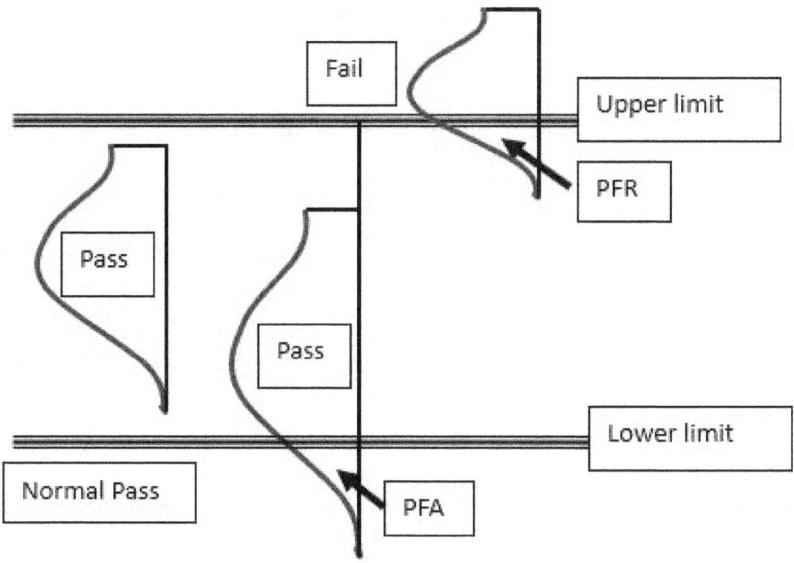

Figure 4-1. An example of Pass and Fail decisions based on a predefined value for PFA and PFR (see text). The probability of false acceptance based on the position of the average is the portion of the area of the uncertainty density distribution which is out of limits although the

average is within limits. One could reject a compatible outcome, or one could accept a wrong outcome when ignoring uncertainty. We refer to PFA (probability of false accept) and to PFR (probability of false reject).

Figure 4-1 shows examples of PFA and of PFR. Judgment based on the position of the average only may lead to risky decision.

Because the risk assessment decisions are based on estimating the probability part, which is out of tolerance value, the computation is straight foreword when the Monte-Carlo method is used [11]. In this case it is sufficient to find the part of the probability histogram lying out of tolerance. Another way would be to find where the Monte-Carlo histogram crosses PFA or PFR points.

The risk that a customer is willing to take is defined in the contract review if conformity is based on risk assessment. ILAC G8 [13] lists some standard risk levels.

Suggestions of PFA levels are:

- NCSL Z540.3 states risky acceptance if PFA>2%, [14].

- The earlier ILAC G8:2009 states PFA<2.5% based on 2.5% on each side of the probability tail when using 95% uncertainty [13]. This is repeated in the 2019 edition.
- ISO 14252-1:2017 states PFA<5% [15].
- Six Sigma concept defines 1 ppm (one part per million) as risky.
- Customer defined in contract review.

4.5.   Conformity method using guard-band.

When using the guard band method for conformity assessment, it is assumed that there is an Acceptance Band narrower than the limits band. All outcomes falling within the Acceptance Band are accepted. The difference between the two is the guard band. This is shown in Figure 4-2.

Using guard band gives the impression that the uncertainty is not involved, however the guard band is usually derived from the expanded uncertainty.

Defining a guard band as in Figure 4-2 can be devised to be equivalent to having the measurement probability density function inside the acceptance band at 95% coverage, at least.

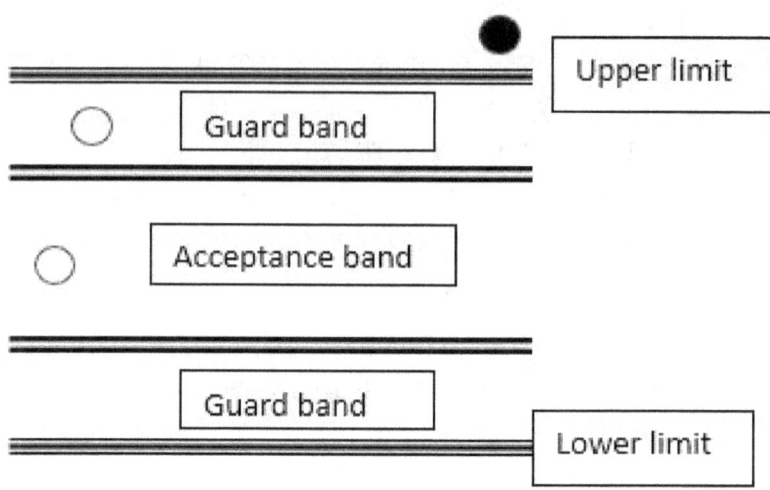

Figure 4-2. Relations between specification limits, acceptance band and guard band. Measurement falling whiting the acceptance band are considered as passed (empty circle).

There are two decision methods without considering uncertainty directly, a binary and a non-binary. A binary decision follows the rule:

- The outcome is anywhere within the tolerance band (empty circles in Figure 4-2) – Pass.
- The outcome is outside the tolerance band (black circle in Figure 4-2) - Fail.

When not considering uncertainty directly, according to the non-binary decision method, there are four possibilities for Pass/Fail:

- The outcome is within acceptance band – 'Pass.'
- The outcome is within guard band and withing tolerance limits – 'Conditional Pass.'
- The outcome is outside the tolerance limits but not farther than the width of the guard band – 'Conditional Fail.'
- The outcome is out of the tolerance band plus the guard band – 'Fail.'

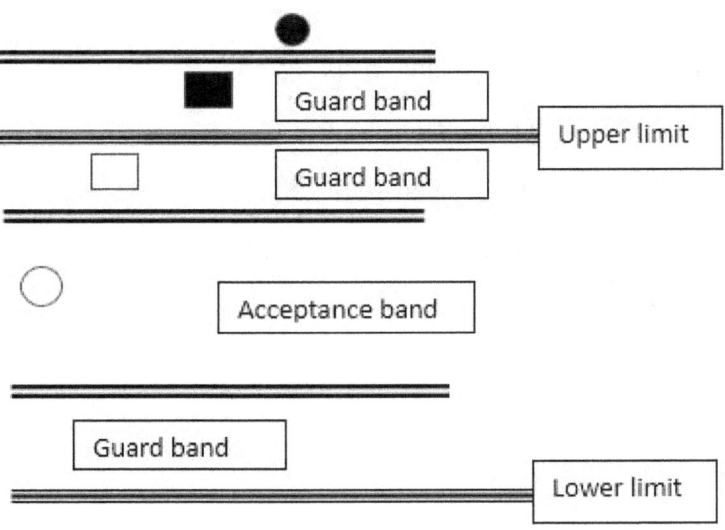

Figure 4-3. Nonbinary case. Black circle = Fail; Black rectangle = Conditional Fail; White rectangle = Conditional Pass; White Circle = Pass. A mirrored situation applies for the lower limits.

Figure 4-3 summarizes the case graphically.

Even if the initial purpose of the guard band was to not deal directly with uncertainties by replacing them, the guard band width does not have to be the same as the expanded uncertainty and several options are described in [12] which correspond to the extent of risks levels proposed in clause 4-2 above. The guard bands (GB) are (U stands for expanded uncertainty):

- Width of 3U equivalent to six sigma. GB=6σ.
- ILAC G8:2009 equivalent to 1U. GB=U.
- ISO 14253-1:2017 equivalent to 0.83U. GB=0.83U.
- Simple acceptance equivalent to 0U and binary decision.
- Customer defined in contract review.

4.6.   Conformity method using uncertainty directly.

When measurement uncertainty is considered without a reference to the guard band, one may decide, as was defined in the earlier edition of ILAC G8, that is, ILAC G8:2007, that:

- If the outcome plus its expanded uncertainty is out of specification limit, it is Fail.
- If the outcome plus its expanded uncertainty is within specification limit, it is Pass.
- Everything else is No Conclusion

Figure 4-4 describes the possibilities. The parallelograms describe the expanded uncertainty and the horizontal lines their outcomes.

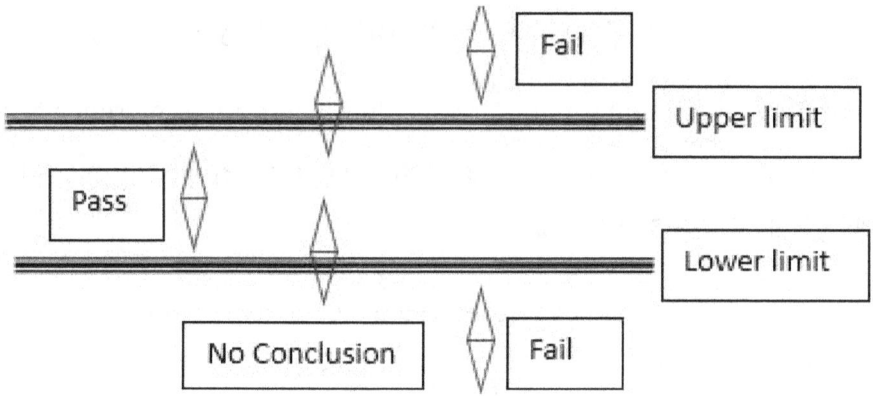

Figure 4-4. Decision considering measurement uncertainty.

5.   Recalibration Interval and prediction

Instruments change overtime. Their uncertainty and measurement error are changing. So, to trust the instrument as measuring correctly one must calibrate it periodically. Also, to be able to use an instrument as a reference, one must be able to predict its value for the moment it is used (normally the time of calibration) based on its past calibrations. To estimate the required time interval, we need to be able to predict how much the instrument changes when used normally and when it will approach its specified limits. Sometimes the changes depend on the number of uses rather than the elapsed time.

In calibration we use reference instruments. We use their uncertainty and measurement error to estimate our UUT's uncertainty and error. The reference instrument also changes over time but what is available to us is an old (maybe one year or more) calibration certificate or certificates. We then need a method to predict the uncertainty and measurement error of the reference instrument to the present time when the calibration is done.

The possibility of being able to predict an instrument with minimal prediction uncertainty depends on the type of instability it undergoes. If the instability is of white noise (meaning, it can change without regard to earlier state up to some extent) the best method to predict is by linear regression. Such a regression supplies the best prediction error and prediction uncertainty. If this is estimated for the reference instrument, then the prediction uncertainty and the error (deviation from nominal) should be inserted in the uncertainty budget (see figure 3-5) as influence quantities since they affect the calibration results. See [16] for more methods of prediction of unstable measurands or instruments.

Due to these concerns, we will describe the prediction process in more detail in the next clause.

5.1. How to predict a future value of an instrument

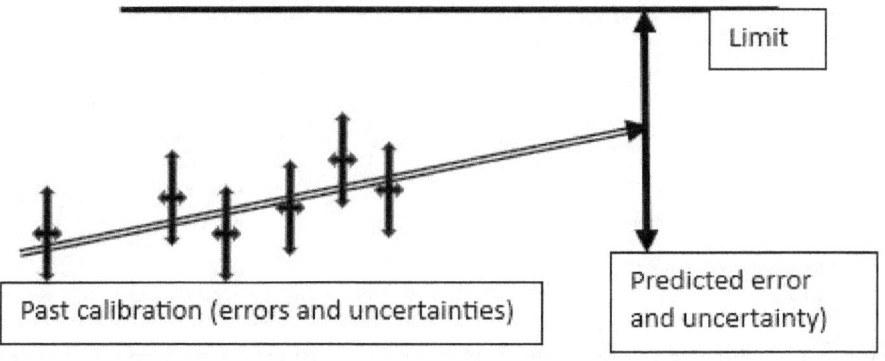

Figure 5-1. If a laboratory has access to past calibrations, a procedure can be devised for prediction. If the past calibrations vary as white noise the best method for prediction is to use linear regression. This allows us to estimate the predicted error and uncertainty and to use them in the uncertainty budget as the reference instrument input quantity [16]. Figures 5-2 and 5-3 show an example of the procedure.

File	Edit	Search	▓Windows	Schedule	Predict	Graph	Report	? Help

🗗🗗 🗗🗗 🔲 ••▷• ••▷• 🖳 ±± ⟳ ⊗stop | **Reload this file** ☐ Enable Editing the Table ☑ Correct for adj.

C:\C++ projects\Develop MetroVal Embarcadero - CX10_3 4-9-2\product\Examples\example.hst

F ᴬ/ₐ ↔ **+Row**   Description: Uncertainties are with a k factor of 2,

Row	Year	Month	Day	Unit1 (for i	Input Value (re	Output Value (	Unit2 (for (	Deviation (ou	Uncertainty (	Parameter (c	Status	ID String (
1	1991	8	27	ohm	10	9.99991	Same	-9E-05	1E-05	DC	old	Tinsley
2	1993	1	27	ohm	10	9.999923	Same	-7.7E-05	5E-06	DC	OK	Tinsley
3	1994	9	4	ohm	10	9.999921	Same	-7.9E-05	5E-06	DC	OK	Tinsley
4	1995	9	4	ohm	10	9.999925	Same	-7.5E-05	5E-06	DC	OK	Tinsley
5	1996	11	9	ohm	10	9.999926	Same	-7.4E-05	5E-06	DC	OK	Tinsley
6	2000	9	12	ohm	10	9.999924	Same	-7.6E-05	5E-06	DC	OK	Tinsley
7	1988	4	21	H	10	9.9999070	Same	-9.30E-05	1E-07	DC	OK	
8	1991	8	27	H	10	9.9999100	Same	-9.00E-05	1E-07	DC	OK	
9	1993	1	27	H	10	9.9999230	Same	-7.70E-05	1E-07	DC	OK	

Figure 5-2. A database of past calibrations. It must include the dates, errors (deviations), uncertainties, information about adjustment, and ID parameters to do prediction estimations.

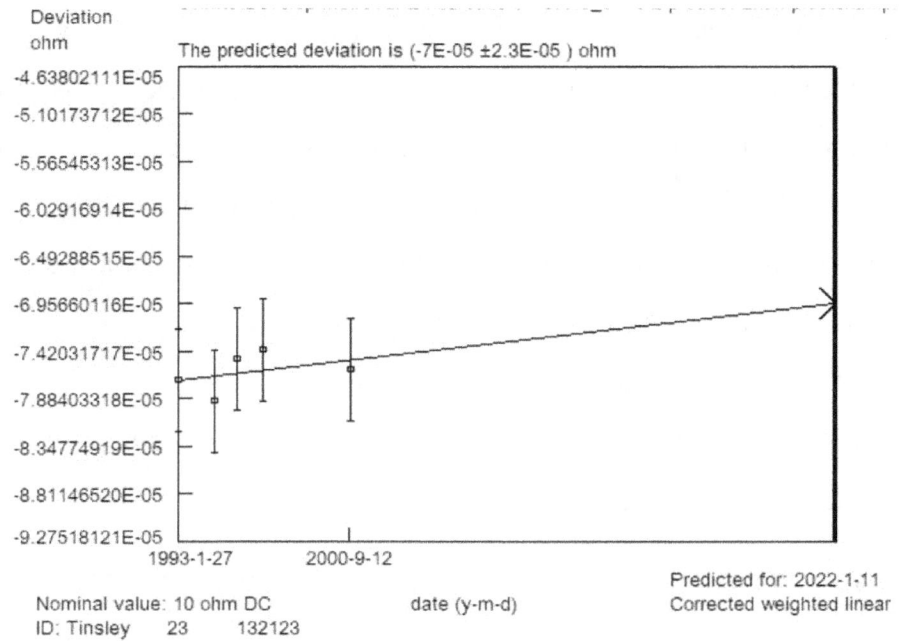

Figure 5-3. A graphical presentation of the prediction of a standard resistor from Figure 5-2 using a weighted linear regression with corrections for the resistor adjustments (no adjustments were required).

5.2. Estimating interval for a specific instrument

The prediction method mentioned above in clause 5.1 is suitable for the estimation of intervals for each instrument separately. To find intervals for all the lab's instruments, the program must run in a loop over all instruments in the database. Figure 5-1 shows the prediction of an instrument at the point when the error plus the uncertainty touches the specification limit. It makes sense to recalibrate that instrument when such a situation is achieved or earlier. To find this date, the software predicts for different dates till it reaches that point and reports the interval. Postponing calibration after that date may result in instruments being out of specification.

Some users do not care about reaching a predefined measurement error since an error can always be adjusted and therefore define an interval as the time it takes for the uncertainty only to reach a predefined value.

5.3. Interval for a group of instruments of same type

The prediction and interval estimations, although done automatically by software, may seem to consume too many resources when cheap instruments are to be calibrated. In such a case the strategy can be to do a batch estimation of the interval on all instruments of a specific model, environmental conditions, and application. The criterion would be to allow for a predefined percentage of the instruments to be out of tolerance. Such a percentage defines the reliability of the instruments. In other words, the reliability of the instruments is here the controlling parameter, and the interval is changed by the program for all instruments till the target reliability is achieved.

6.   Bibliography

[1] Temple and Williams (2002) in the report "The Economics of Metrology and Measurement", GM Peter Swann, Final draft 14th October 2009.

 [2] Peter J Mohr et al 2018 Metrologia 55 125

 [3] 26th CGPM Resolutions on the revision on the International System of units (SI), 2018.

[4] https://www.bipm.org/kcdb/

[5] A Lepek et al, Metrologia 1995/6, 32, 245-252

[6] JCGM_100_2008.pdf, Evaluation of measurement data - Guide to the expression of uncertainty in measurement. May be downloaded from https://www.bipm.org/.

[7] JCGM_101_2008.pdf, Evaluation of measurement data – Supplement 1 to the Guide to the expression of uncertainty in measurements – propagation of distributions using Monte-Carlo method. May be downloaded from https://www.bipm.org/.

[8] JCGM_102_2011.pdf, Evaluation of measurement data – Supplement 1 to the Guide to the expression of uncertainty in measurements – extension to any number of output quantities. May be downloaded from https://www.bipm.org/.

[9] Statistical distributions, Meran Evans et al., John Wiley and sons, 2000

[10] Ambient air – Standard method for the measurement of the concentration of Sulphur Dioxide by ultraviolet fluorescence; English version EN 14212:2012, English translation of DIN EN 14212:2012-11, November 2012

[11] MetroVal, calibration and metrology analysis software, https://www.newtonmetrology.com

[12] Lepek A, NCSLI conference proceedings, 2012, The implementation of supplement 2 to the GUM in software.

[13] ILAC -G8:09/2019, Guidelines on Decision Rules and Statement of Conformity

[14] https://ncsli.org/page/z5403

[15] https://www.iso.org/standard/70137.html

[16] A Lepek, Clock prediction and characterization, Metrologia 1997, 34, 379-386

[17] Lepek A, Software for the prediction of measurement standards, NCSL International Conference, 2001

[18] ILAC -G24 and OIML D 10, Guidelines for the determination of recalibration intervals of measuring equipment, 2007. Method 5, based on [17].

www.ingramcontent.com/pod-product-compliance
Lightning Source LLC
Chambersburg PA
CBHW072335290526
45794CB00002B/888